Artificial Intelligence and the Christian

Understanding AI's Promises and Pitfalls

Peter Goeman

SOJOURNER
PRESS

Artificial Intelligence and the Christian: Understanding AI's Promises and Pitfalls

Copyright © 2024 by Peter Goeman

Unless otherwise indicated, all Scripture quotations are taken from The ESV® Bible (The Holy Bible, English Standard Version®) copyright © 2001 by Crossway, a publishing ministry of Good News Publishers. ESV® Text Edition: 2016. The ESV® text has been reproduced in cooperation with and by permission of Good News Publishers. Unauthorized reproduction of this publication is prohibited. All rights reserved.

Cover design by Alam

Cover photo by monsitj / Double exposure image of virtual human 3dillustration stock photo / iStockPhoto (id: 695144104)

ISBN 978-1-960255-14-3 (Paperback)
ISBN 978-1-960255-15-0 (Epub)

Printed in the United States of America
Sojourner Press
Raleigh, NC
sojournerpress.org

For bulk, special sales, or ministry purchases, please contact us at sales@sojournerpress.org.

Contents

Preface	1
1. Don't Trust Your Eyes!	3
2. The Power and Possibilities of AI	7
3. The Dark Side of AI	23
4. Biblical Guidance for the AI Age	43
Conclusion	54
Other Books by Sojourner Press	56

Preface

Although Solomon's words are true—"There is nothing new under the sun"—that does not mean we do not face new technological challenges. Artificial intelligence (AI) presents Christians with a unique challenge that few have fully understood or even thought through. My goal in this book is not to exhaustively analyze the ethics of using AI or dive deeply into the technicalities of how AI works. Rather, this book is designed to be informative. I recognize the pressing need for Christians to be instructed not only in what AI is capable of but also in the real dangers that AI innovation poses.

I wrote this book in response to multiple requests from pastors and church leaders who wanted presentations on AI and practical instruction on its potential benefits and risks. As we live in an unprecedented time of technological innovation, Christians must be equipped with wisdom to think carefully about the issues AI introduces. It is my hope that this book will serve as a guide to help believers engage with AI thoughtfully, applying biblical principles to this modern tool.

AI can either serve as a powerful tool for good or become a source of harm, depending on how it is used. I hope that through this book, you'll gain the insight and discernment needed to navigate its use, but also be wary of the dangers.

Furthermore, I pray this book will stimulate discernment and deeper reflection on how we, as followers of Christ, can benefit from AI without losing sight of our values and convictions as Christians. May we approach AI not with fear, but with caution, endeavoring to glorify God in all things.

Peter Goeman
September, 2024

Chapter One

Don't Trust Your Eyes!

The winner of the 2023 Sony World Photography Awards was Boris Eldagsen. The photo was black and white, featuring an older lady hugging a middle-aged lady from behind. It was beautiful. It looked old, and it was the recipient of the $5000 first-place prize. The only problem? It wasn't real.

Eldagsen shocked many within the photography world when he turned down the coveted prize and revealed that he had used Artificial Intelligence (AI) to create the image.[1] He had actually not taken a picture at all. He had edited and re-edited the photo 20 to 40 times through image generators. He said that his goal in being a "cheeky monkey" was to bring awareness to the fact that generative AI should not compete against photography because they are different entities. But ironically, his AI generation was so much like photography that it won the award for best photograph for the open category! The judges had not been able to discern the extent that AI had been used in the photo by Eldagsen.

---

1. Paul Glynn, "Sony World Photography Award 2023: Winner Refuses Award after Revealing AI Creation," *BBC*, last modified April 18, 2023, https://www.bbc.com/news/entertainment-arts-65296763.

When I first began studying and presenting on AI, I often had to demonstrate what generative AI was capable of. To show people the full extent and capability of AI, I would often include pictures that looked real, but were clearly historical fabrications. I would include photos of President Trump sitting down for a picture with Winston Churchill, a photo of Trump winning the 1936 Olympics, and an image of Trump singing alongside Taylor Swift. Not playing political favorites, I would also include videos of President Obama advertising for KFC, sitting down with President Putin and eating some tasty grilled chicken. None of these things ever took place in real life, but with generative AI, it looked like it had happened. And all of this was done with free tools available to anyone with an internet connection or a smartphone.

It often blows peoples' minds when they realize what AI is capable of. For the last year and a half, various AI programs have shown themselves capable of generating almost any kind of image, regardless of whether it happened. Furthermore, as of the writing of this book in 2024, AI is continuing to make massive innovations in generating not just pictures but also movies and videos.

In 2016 AI technology was used to present Carrie Fisher (Princess Leia) and Peter Cushing (Governor Tarkin) as living characters in *Rogue One: A Star Wars Story*.[2] Through painstaking effort and significant cost, Leia and Tarkin were pictured in the 2016 film exactly as they had been in the

---

2. Samit Sarkar, "Rogue One Filmmakers Explain How They Digitally Recreated Two Characters," *Polygon*, December 27, 2016, https://www.polygon.com/2016/12/27/14092060/rogue-one-star-wars-grand-moff-tarkin-princess-leia.

original 1977 film, *Episode IV: A New Hope*. The audience was blown away at seeing Tarkin and Leia again. If you watch the movie, it is quite well done and most people can't tell it was computer generated. But within the last eight years you won't be surprised to know that the technology has grown by leaps and bounds. It truly is either spectacular or terribly frightening (depending on your perspective), or perhaps a little of both.

In the past, we used to have a tongue-in-cheek rule: "Take pictures, or it didn't happen." Now we live in a day where, even if it didn't happen, there will be pictures! It is nearly impossible for the average individual to discern between a real photo and a doctored AI photo.

Although the visual capabilities of AI are quite stunning, the text composition capacity of AI is also worthy of note. AI language models, like those used in chatbots or text generators, are able to produce eerily human-like responses, articles, and stories. These models can even mimic specific writing styles or generate poetry that resonates emotionally with readers. However, as with image generation, there are dangers lurking within the realm of AI-generated text. As AI becomes more adept at mimicking actual writing, it becomes harder to distinguish between what is authentically human and what is machine-made. This can blur the line between truth and fabrication in written communication, posing significant challenges for Christians, who value discernment and integrity.

Such technological innovations are exciting, to be sure. Movies like Star Wars and Indiana Jones have used AI to recreate past characters in different scenes, and we certainly enjoy that. But we also suffer from the evil use of this kind of technology. Images and videos that portray events and scenarios that never actually existed will certainly become more and more common.

Social media and the Internet had already posed problems for Christians in being discerning about what was true and what was false or misleading. With the innovations of AI, believers are going to have to work twice as hard to be discerning and to ascertain the truth.

Without a doubt, AI has brought with it amazing possibilities. At the same time, as you can clearly see, there are also tremendous dangers involved with this. It is not my goal to solve all of these issues for you. But it is my modest goal to take the next few chapters and walk you through ways that Christians should think through AI. We should heartily acknowledge that Christians can benefit from AI. But we also need to warn Christians about the potential dangers of AI. And perhaps more important than anything, we need to apply biblical principles to technological advances and understand our role as being made in the image of God.

Chapter Two

The Power and Possibilities of AI

It would be a mistake to run away from AI and avoid using it at all costs. In fact, it would technically be impossible to do that! AI has already been incorporated into your life, even without you being aware of it. Your bank account uses a form of AI to monitor transactions, your smartphone uses AI in multiple ways, including suggesting the next word of your text. Netflix, Google, and Amazon all incorporate AI into their algorithms and programs.

Although it's impossible to avoid using AI, it can be beneficial to use AI intentionally and intelligently to help serve the Lord more effectively and efficiently. Below are a few ways I think Christians will be able to benefit from AI in the coming months and years. The list will undoubtedly expand and change as AI grows and develops. But these are helpful starting points.

Personalized Learning

Are you interested in learning a new subject? How would you start learning about it? What books would you read? What would be your pace of study? The most common difficulty in starting to learn a new subject is knowing where to start. One of the main reasons people do not learn new things is because they do not start. But imagine if you could go to a website and simply type in a request to create a customized learning plan for a particular subject. Your request could include a study plan along with a suggested reading.

This is already possible with many AI models. You can request custom learning plans for Spanish, French, Chinese, etc. You can ask for custom learning schedules related to subjects like math, science, and history. Since many of us struggle with knowing where to start or how to progress, it can be effective to have AI design a learning plan and schedule for a subject that we know very little about.

AI has a presumable advantage because it has been trained on significant material in any given subject field. I recently heard one well-known educator say that AI could conceivably eliminate the need for public school. Parents who have refused to homeschool their kids because they were unable to plan lessons effectively now have a cost-effective AI administrative assistant to help.

Grammar, Spelling, Rewrites

If you've ever used some sort of grammar check or spelling aid like Grammarly or ProWritingAid, or any of the other numerous programs that are out there to assist your grammar and spelling, you've likely already been utilizing AI. Many of the grammar and spellcheck programs now incorporate AI to help with knowing whether or not your grammar is correct and your spelling is accurate.

This is a superb benefit for those of us who want to write as clearly and succinctly as possible. For example, I wrote this book by dictating a significant portion of it into my Iphone. Then, I fed that vocal transcription through AI to correct any obvious mistakes and to ensure that what I had spoken would read clearly. I made many of the suggested corrections that AI gave me, and then I read it through on my own to double-check everything. I also had an external editor then look it over. The point is that AI was quite helpful in producing the rough draft of this book.

One way I often use AI is to provide suggestions for rewriting a sentence for clarity. For example, sometimes I'm trying to say something specific, but the sentence does not read how I want it to. Perhaps I can't think of the right word, the grammar just doesn't sound right to me, or maybe other factors are involved. In such cases, I will often copy that sentence into a popular AI program like ChatGPT and ask the program for ten alternative ways to write that sentence. Usually, I end up combining three or four sentences into one unique sentence to express the statement exactly how I want it to be said. This helps provide clarity and helps me as the author understand whether I am writing in a way that can be understood. I have occasionally submitted a sentence to ChatGPT or other AI programs and quickly realized that if AI didn't know what I was trying to say, nobody else would either. Back to the keyboard on those sentences!

Brainstorming Ideas

One of the most profitable ways to use AI is to brainstorm ideas. Perhaps you're not sure what to do for the upcoming men's retreat. You can ask AI to give you 20 different ideas to do at a men's retreat for a church. Sure enough, you'll have lots of suggestions. Interestingly, what often happens when I use AI to brainstorm is that the ideas themselves may not be useful to me, but by reading those different ideas, my mind comes up with a different idea. So, although AI itself did not create the idea I used, it was the indirect source for that idea by priming the pump.

I talked with one lady at our church who used AI to come up with the name of her company. She asked ChatGPT to give her 50 different possibilities for the name of a company specializing in firearm training. She said that one of the names was perfect for her, so she used it and incorporated it into an LLC.

Pastors planning on interviewing one of the church's missionaries may want to brainstorm different questions related to the missionary's experience. The pastor could easily ask AI to provide 15 possible questions for a missionary living in northern Africa. This kind of brainstorming capacity of AI can help save time and promote other ideas in and of itself. The brainstorming capability of AI is undoubtedly one of the most beneficial ways a Christian can use AI.

Event Planning

I am not, by nature, an event planner. I *can* plan an event, but I don't necessarily enjoy the process. I would much rather use my time in other ways, like spending it with my wife and kids or studying the Bible. However, it is necessary to plan events. Thankfully, there are often gifted individuals within churches who thrive in positions where they organize the logistics behind events. Praise God for those individuals! But sometimes, for small Christian companies and churches, there is no one who can plan events effectively. In such cases, AI shows promise in helping those who, like myself, are not necessarily inclined to plan important events.

Due to its training, AI is highly capable of helping organize events and schedules. For example, perhaps a church might need to create a schedule for the VBS program. In such a case, we would want specific ideas to be implemented, but there would also be flexibility about when to implement them. To help us organize such a schedule, we can ask AI to design a schedule within those parameters. We may not embrace the final product, but we can edit the suggested schedule and save you time (theoretically at least).

AI can help with many similar events. Consider the volunteer who wants to organize a group of small group leaders, a music team, a setup team, etc. AI can help schedule those kinds of situations and even help evaluate whether or not something might be missing.

Audio, Picture, and Video Editing

One of the exciting innovations that has taken the technological world by storm is the capacity for AI to edit audio, pictures, and videos. I do not consider myself an expert in the world of audio and video editing, but I have had to learn quite a bit on my own thanks to running The Bible Sojourner podcast and YouTube channel. I actually enjoy the process, but it's very time-consuming. So for efficiency's sake, one of the most exciting aspects of AI is how quickly it can edit an audio file, which would potentially take much longer to edit. As of the writing of this book, Adobe has a beta AI audio program that is available for free, and you can upload an audio file to their website.[1] The AI will then edit and clean up the audio file, adjusting the levels appropriately and removing any unnecessary background noise. It does this without you having to do anything. Granted, AI is still in the beginning stages of being able to do this, but I have used such programs myself multiple times and have been delighted to see what it is capable of.

Concerning image editing, I am essentially a fish out of water. I have no artistic ability whatsoever. However, with AI tools, a graphically challenged individual like me can type a prompt into a program like Adobe Photoshop or Canva, and AI can work on a photo, adjusting it according to any request. It is important to remember that these tools are just now emerging on the market. Much of these developments have come within the last two years. Imagine what might be possible if things continue to progress at this pace for another five years!

Producing Royalty-Free Images

Along the same lines as the previous discussion, AI is not only capable of editing images but can also generate images that do not exist elsewhere. In other words, AI can produce or create a unique image that does not have a license restriction attached

---

1. https://podcast.adobe.com/

to it. And depending on the type of image, AI can create some very realistic pictures.

This is useful for Christians in a variety of capacities. One way that Christians might benefit from generative AI is to use AI images for certain church communications. For example, suppose you're planning a church picnic and you're worried about using an image with a royalty fee attached to it. Typically, you need to purchase the license or rights to use a stock photo or image. There are free stock photo websites, and the price for using royalty images is usually not high. However, depending on the budget, it can be inconvenient or not an option. In such cases, AI can generate an animation, cartoon, or realistic photo of a family or a church group at a picnic alongside the riverbank. You can make the image as specific or as general as you want. And if you aren't satisfied with what the AI image generator created, you can simply regenerate the image. I often use an image generator for my blog posts and some of my YouTube thumbnails. One of my colleagues at the seminary uses AI photos for his PowerPoint presentations almost exclusively.

I will say that, unsurprisingly, some lawsuits are pending on whether or not AI and its generative image creation should be allowed or how to regulate it. Those lawsuits will likely be in the courts for years to come. There may be laws in the coming years restricting AI image creation, but for now, there are no restrictions on using images created by AI.

Research Assistant

In the past, to find out if a book or article dealt with the subject you were studying, you would either have to read it or spend significant time skimming through sections of the work yourself. Now, you can ask AI to give you a synopsis of an article or book. You can also upload a specific article as a PDF and have live interaction with an AI research assistant, asking it particular questions, such as whether this article talks about Matthew 16, refers to the Ascension of Christ, or deals in depth with the baptism of the dead. I often use AI to understand if an article or book deals with what I am looking for so that I don't spend precious time looking through an article or book that is irrelevant to the topic I'm trying to research.

Additionally, AI has made search capabilities a lot more specific and functional. Long gone are the days of keyword searches. As AI continues to revamp search capacity, one can search journal databases and libraries with more and more specificity and exactness.

AI cannot replace spending time grappling with ideas and reading the material for oneself. But it can certainly save precious time for the individual who wants to spend that time reading and thinking rather than looking for resources. AI's capability as a research assistant is going to continue to grow, and it seems prudent to take advantage of such a tool.

Translation Help

One of the areas I'm most excited about in AI innovation is language translation. AI large language models (LLM) are specifically trained on large data sets from multiple languages. The advantage of this kind of training for AI is the ability to recognize what words and phrases most naturally go together in specific contexts. This kind of translation approach is very profitable.

To illustrate how this works, and why it is so beneficial, consider the fact that when people are learning a new language, they often mistakenly think that each word in their native language has a corresponding word in the language they are trying to learn. But that's not how language works. Language works together in phrases, sentences, and paragraphs, which together form a specific context. It's rare that a word-for-word translation equivalent is always workable. Large language models often recognize this and will translate much more realistically than some of the translation programs that have existed in the past. Although this technology is in its initial phase, there have already been many encouraging developments in how AI recognizes the functionality of language and can translate that into other target languages.

The uses of this technology are various for the Christian. For example, consider you are a missionary who wants to go to a foreign context, learn a new language, and translate the Bible into that language. Historically, you would have to spend significant time cataloging language uses and even creating lexicons or dictionaries for the target language. Although all of that work certainly is not in vain, AI opens up new possibilities. In such a situation, AI could record many audio conversations in a tribal context and transcribe them, analyzing word usage, and context. With enough data, AI could presumably construct a rough lexicon or dictionary for the target language, giving a valuable starting point for the Bible translator. However, AI is not without error and it would be imprudent, to say the least, to entrust such a valuable process to a machine. So one would need to double check what AI has done. But it is often much more useful to have a rough starting point than to have to start a project like that from scratch. In any case, it would be foolish to ignore that there is a tool in AI that allows in-depth language analysis that was previously unheard of.

Mundane Automation

AI is well known for its ability to handle tasks that require mundane automation. Prior to AI, computers were already very capable of handling tasks that were the same every day. With the advances in AI, the ability to handle mundane automation has increased. Companies are currently making innovative strides in all sorts of possible applications to automation. Several companies are promoting self-driving vehicles powered by AI, though one might question whether driving truly qualifies as a mundane task. Meanwhile, AI has rapidly become integrated into more routine, everyday activities. AI has been used to schedule interviews, haircuts, office hour appointments, etc. It appears that AI will automate many tasks that don't require extensive thought. While AI hasn't fully reached its potential yet, we are still in the early stages of its development.

Medical diagnostics

One area where AI holds significant potential is in medicine and healthcare. According to Business Insider, in early 2023, ChatGPT passed the US medical licensing exam and didn't just scrape by! In many ways, ChatGPT performed better than a lot of doctors. Apparently, for example, when taking the medical exam, it diagnosed a rare condition many doctors often miss.[2] One of the reasons AI can thrive in the situation of medical diagnostics is that it is completely emotionless and functions entirely on logic and parameters that have been programmed into it. I have personally asked different AI programs medical questions about my family, friends, and even myself, and I have been impressed with the medical advice and knowledge given.

As an example of the capability of AI in diagnosing medical problems, I read recently about an AI program that can use the smartphone camera and diagnose a stroke with 82% accuracy![3] Just by holding up the camera to your face, you can have a fairly good chance of diagnosing a stroke correctly! When minutes count in treating a stroke, this technology could be life-saving. This is just one example of the exciting advancements in the medical world due to AI. And it seems we are just beginning.

Now, it goes without saying (yet I am obligated to say it anyway) that AI is no replacement for a real doctor. However, I can see great benefit from AI to aid medical diagnostics, perhaps even as an aid to the triage process in the emergency room. In a day when we continue to learn more and more information, it can be an effective tool to have AI assimilate

---

2. Hilary Brueck, "The Newest Version of ChatGPT Passed the US Medical Licensing Exam with Flying Colors — and Diagnosed a 1 in 100,000 Condition in Seconds," *Business Insider*, last modified April 6, 2023, https://www.businessinsider.com/chatgpt-passes-medical-exam-diagnoses-rare-condition-2023-4?op=1.

3. Sujita Sinha, "AI: Phone App Detects Strokes from Face in Seconds with 82% Accuracy," *Interesting Engineering*, June 20, 2024, https://interestingengineering.com/innovation/ai-app-on-smartphones-spot-strokes.

all that data and compare and contrast it with a patient's symptoms. A human doctor has limited time and ability to consume all the information on a given subject. This area of AI innovation has great promise indeed!

Financial Fraud Monitoring

In a digital world, it is impossible for human beings to keep track of every transaction and purchase. Many banks have already been implementing advanced fraud detection services. Perhaps you've gotten a notification on your phone asking if you had made a certain purchase. AI has the potential to greatly improve such financial monitoring. Although one person can barely monitor a handful of transactions in real time, AI can handle millions of data points, spotting anomalies that would indicate fraud or theft.

Concise Summarization

It is an ill-kept secret that many people like to sound smarter than they are. Unfortunately, that often comes through in how they write. For some reason, humans like to use long, difficult-to-understand words or technical jargon, making it difficult for the average person to understand what is being said. This happens in theological works, medical assessments, contracts, and even your HOA agreement. AI allows everyone the opportunity to scan a document and get a concise summary of what is being said. Not only will it rephrase things to help you understand, but it will also provide you the opportunity to interact and ask questions about specific parts of documents that need further summary. AI's ability to summarize long and complex documents is certainly an attractive and useful feature.

Reading Books and Documents

If you've ever used the original Microsoft automated reader, you're familiar with the cringe-worthy experience of having a computer read a text file to you. In the past, these computerized readers would make various mistakes in pronunciation or timing. There would be mispronounced common words, too little spacing between words, or sometimes too much spacing between words! In the past, I've tried multiple times to give my eyes a break from reading on a screen by listening to the automated reader. But it always ended in disaster and discouragement, not being useful in any capacity for me.

Today, thanks to AI, automated reading has undergone vast improvement, and in many cases, it sounds equivalent to a human reading a text. These technological advances have great promise in a variety of uses. For one, virtually every book ever written now can be made into an audiobook. Those who are blind or visually impaired can benefit from programs that will automatically convert a book to audio. In many cases, AI also provides the ability to choose from different voices. In fact, technology already exists for an AI voice to imitate anyone who has significant amounts of audio material available online. This includes celebrities, pastors, politicians, etc. Can you imagine choosing to listen to your audiobooks in the voice of your favorite actor or actress?

For people who do a lot of driving, there are now opportunities to listen to blog or journal articles they would not previously be able to read due to time constraints. Where books previously required a human narrator to record and edit in a lengthy process, now, with a push of a button, any text can be read in virtually any voice.

This kind of technology is not limited to the commercial industry. I recently had a conversation with a nonprofit company working on a partnership with an AI company to make Bibles available in audio format with multiple voice selections. I personally would definitely sign up to listen to the Bible read by Alistair Begg.

Email Filtering and Replies

More of our communication happens by email than ever before. It shows no sign of stopping. There are emails from the grocery store, church, politicians, your favorite blogs and podcasts (like the Bible Sojourner), not to mention emails from friends, family, work, etc. One particular benefit of AI is that it can function like an administrative assistant. I think I get a lot of emails, but I know some of you get way more emails than I do! AI essentially provides an unpaid staff position to help you organize and respond to emails.

Being a seminary professor, I get a lot of interview requests and speaking opportunities. I cannot take on all of them, but I also don't want to ignore people who take the time to contact me. In the past, I might have used a "template of rejection"—that sounds loving, doesn't it—and I might have edited a few words and phrases to copy-paste a response in an email saying why I wasn't able to do the interview or come speak at a church. But now, I can actually have AI craft a response based on how I would typically respond. AI is creative and unique, using real details from the email request in its response. As I look over the email before sending it, I will often change some words or phrases I don't particularly appreciate. I might also add a sentence or two that I think would be appropriate. But overall, it saves me a lot of composition time to write what I would write anyway. I am a much better editor than I am a writer concerning those kinds of emails.

Using AI to handle some of the more mundane email responses is similar to how many pastors have used administrative assistants. In the past, pastors and corporate executives have hired administrative assistants who can craft emails and handle some of the everyday conversations. That is a valuable task, but not everyone can afford an administrative assistant! AI can help with this aspect of life. For those Christians who receive a lot of emails but can't afford an administrative assistant, using AI is one way to increase productivity. And perhaps you might even answer a few emails that you otherwise wouldn't be motivated to respond to.

Summary

This chapter has looked at a variety of ways that we can use AI to maximize our productivity and usefulness to the Lord. As Christians, we should desire to take advantage of tools that will allow us to serve the Lord better. Even the simple idea of using AI to check spelling and grammar goes a long way in representing Christ well in how we communicate.

It is undoubtedly true that there are many advantages and benefits to using AI. However, even as we discussed some of the benefits and advantages, it probably crossed your mind that there are potential dangers built into some of these potential advantages. To be sure, significant potential dangers are involved with AI advancements. Just like any tool, there are positive and negative uses. The same dynamite that can blow a valley through a mountain to make way for a road can also destroy a bridge that connects two major cities. The same hammer that can help build a house can destroy and even kill. There is a reason some people view AI as the solution to all of humanity's problems and also the inevitable cause of its destruction. In the next chapter, we will take a serious look at some of the dangers of AI.

Chapter Three

The Dark Side of AI

Many people view the recent innovations in AI as the salvation of mankind. AI will solve world hunger, cure all diseases, and solve all strife in the world! But AI is simply a tool. It's a creation of mankind. Although we have seen in the prior chapter that AI can potentially produce some fantastic benefits, we need to be aware of some significant dangers embedded within the technology.

This chapter focuses on the dangers and potential problems that AI will introduce or aggravate for society, particularly impacting Christians. Christians and Christian leaders especially need to be aware of these dangers. A new world is developing around us, and we need to be prepared to live in it.

Phishing and Deception

We've all received those emails from a "Saudi Arabian prince" claiming to be in dire straits, yet promising to share his fortune of millions with you if only you can lend him a few thousand dollars. All he needs is your bank information to make the necessary withdrawals, after which he'll deposit millions into your account! The appeal of these emails is baffling, especially given their frequent errors and misspellings. Yet, surprisingly, these phishing scams do succeed. In fact, the Federal Trade Commission reported that phishing scams resulted in a loss of $52 million in 2022.[1] Not bad for a prince who can't spell very well!

In the previous chapter, I highlighted how AI has introduced numerous helpful tools for checking spelling and grammar. AI can even help compose short messages for you in a professional tone, a casual voice, or with a touch of humor. What's more, if you have a sample of a communication style you want to emulate, AI can easily accomplish that. For example, if someone fed this book you are reading into an AI system for analysis, it would be possible to produce a message or a similar book in a style that mimics mine. AI would use similar descriptions, phrases, and vocabulary, making it read like my writing. I'm not even sure my mother could tell the difference!

Now imagine Grandma Betsy, who gets an email from her granddaughter traveling in Europe. Grandma Betsy recognizes some of the communication patterns from her daughter and asks no questions when the daughter asks for a transfer of $2000 to a specific Swiss bank account so that she can purchase return flight tickets home.

Or consider what I mentioned in the last chapter about AI's ability to mimic voice patterns. Several friends of mine have been "deepfaked," a term for when someone's identity is

---

1. Jo Rushton, "50+ Phishing Statistics You Need to Know – Where, Who & What is Targeted," *Techopedia*, March 1, 2024, https://www.techopedia.com/phishing-statistics.

replicated to imitate them. Their grandparents have received calls from a voice that sounds just like my friends, asking for financial assistance. Similarly, a lady in our church received a call from her supposed granddaughter, instructing her to send a check to a certain organization on her behalf. Sadly, that grandmother is now out thousands of dollars.

Unfortunately, older adults are often victims of this kind of fraud and attack. The elderly in churches are especially vulnerable because they naturally want to be gracious, kind, and giving to real needs! It is a low move to take advantage of those who deeply desire to do good and cheat them out of their hard-earned money. But that is what evil people do, and there is a lot of evil in this world.

However, it is not just the elderly who are susceptible to "deep fake" deception. One employee from Hong Kong, a worker in a multinational firm, was duped into paying $25 million to fraudsters who used AI technology on a teleconference call to impersonate the CFO and other employees of the company.[2] Although the employee was initially skeptical of the order to transfer funds (which came by email), he recognized the faces and voices of coworkers and was convinced that this was a legitimate order.

We need to do a better job in our churches of warning everyone, especially the elderly, about the scams and phishing attacks that can come through phone calls, text messages,

---

2. Heather Chen and Kathleen Magramo, "Finance worker pays out $25 million after video call with deepfake 'chief financial officer,'" *CNN*, February 4, 2024, https://www.cnn.com/2024/02/04/asia/deepfake-cfo-scam-hong-kong-intl-hnk/index.html.

and emails. Unfortunately, just because you recognize your grandson's voice and they ask about your dog doesn't mean you're actually talking to your grandson.

Although much more could be said on this issue, here are four brief principles that could help us avoid being scammed.

1. *Verify the request through a separate channel*. If someone contacts you asking for money or personal information, contact them back through a known and trusted method. For example, if you receive a suspicious email from a friend, call them directly to confirm.

2. *Ask personal questions*. To verify someone's identity, ask a question only the real person would know the answer to—something not easily found online or through social media. I once encountered a suspicious social media account pretending to be a pastor friend of mine. To test them, I asked if they remembered our conversation in New York, fully aware that we had never been to New York or discussed anything related to it. The imposter responded enthusiastically, claiming they enjoyed the conversation and were grateful to reconnect. Needless to say, I promptly deleted that fake account and had no further interaction.

3. *Use multi-factor authentication*. For financial transactions or sensitive accounts, enable multi-factor authentication (MFA). This requires an additional step beyond just a password, such as a code sent to your phone. This step makes it all the more difficult to make unauthorized transactions on your behalf.

4. *Consult with someone you trust*. Before making any financial decisions or sharing personal information, discuss the request

with a trusted friend or family member to get their opinion and ensure it's legitimate. This is an area where church leaders and elders can help shepherd the congregation. Shepherds are not just around to feed the sheep. Shepherds often need to function in their role as protectors as well.

Sexual Temptation

Sexual temptation has always been around, but technological advances have introduced new and unique challenges. I often wish I could travel back to the early days of the Internet. I was quite young then and don't remember much, but I wonder if any Christians foresaw that the most visited websites and the most widespread use of the Internet would revolve around pornography. My guess is they had no clue. Now, with the Internet, social media, and other technological advances, we've seen the profound impact technology has on humanity—and unfortunately, much of it isn't good.

Pornographic websites make up the majority of web traffic. Famous porn websites, which I will refuse to name here, are visited over 700 million times more than household names like Amazon and Netflix over the year. Data shows that not only do people visit porn sites more often than other websites, but people also spend more time on those websites.[3] It is a sad and open secret that a significant "side effect" of our technological revolution is pornography.

---

3. Nicole K. McNichols, "How Many People Actually Watch Porn?" *Psychology Today*, September 25, 2023, https://www.psychologytoday.com/gb/blog/everyone-on-top/202309/how-much-porn-do-americans-really-watch.

Technology has undeniably created problems for society, aggravating specific temptations and pitfalls for believers. Yet, we must recognize that technological advances won't alleviate these issues. In fact, with the rise of AI, sexual temptation is poised to intensify. Websites and online ads will become even more adept at tracking your activity, tailoring content to your preferences, and luring you in. What's more, the offerings will become increasingly irresistible, making the battle against temptation more challenging than ever.

The ability of AI to generate images and videos has many positive uses, as we discussed in the last chapter. However, with regard to the dangers, the pornography industry is positioned to benefit significantly from being able to create inappropriate, sinful, yet realistic images and videos. Technology will enable companies and individuals to create images and videos without having to worry about rights, licensing, being sued, or being accused of rape. I hesitate to say it because the sinful imagination is powerful, but this kind of technology really opens up the possibilities of anything you can imagine—sexual fantasies made to order by the click of a mouse or the press of a button.

It seems inevitable that the church will need to counsel a generation conditioned to gratify every desire or imagined thought in the sexual realm. The church must use Scripture to guide individuals back to the foundational understanding of marriage—a selfless and sacrificial relationship for the mutual benefit of both husband and wife to the glory and honor of Christ. Christians, especially the young, must be trained in the foundational principles of marriage and Christian character to protect them from the notion of self-centered sexual pursuit.

Another significant issue in this category is the existence and potential of deep fakes. A deep fake is a digitally manipulated video, audio, or image created using advanced AI techniques to alter or fabricate someone's appearance or voice. Although this can be done in humorous examples, such as showing Uncle Joey fighting a grizzly bear, it can also be used in nefarious ways by showing someone in compromising or evil situations. For instance, in 2023, there were 143,000 deep fake nude photographs shared online.[4] This means that a photograph was edited to make an individual appear as if they were posing naked or engaging in some uncouth act.

This has been particularly devastating for young school-age girls who may not have done anything wrong. Yet, a classmate could use an innocent photograph to create a fake nude image through AI, which is often indistinguishable from real photos. While legal action has been taken against such offenses, no law can curb the sinfulness of the human heart. This is a very real danger in public schools today, and it is hard to imagine these tendencies slowing down in such environments.

I wasn't around when Christians were discussing the arrival of the Internet and its potential problems and dangers. However, witnessing the innovations of AI, it's hard to see it as entirely beneficial. I have a heavy heart, knowing the potential evil that could be unleashed through this technology and the fact that future generations will have to deal with its aftermath.

---

4. Associated Press, "Teen Girls Are Being Victimized by Deepfake Nudes across the Country," *Boston Herald*, December 2, 2023, https://www.bostonherald.com/2023/12/02/teen-girls-are-being-victimized-by-deepfake-nudes-across-the-country/.

Isolation and Loss of Community

In a sad twist of irony, despite all the technological advances and social media innovation, loneliness among the upcoming generations is skyrocketing. Over the last ten years, social media has promoted isolationist tendencies among youth growing up. It is extremely well documented that young people who have grown up with access to social media are suffering from loneliness and isolation.[5] There is no reason to expect that advances in AI will not have a similar effect.

With the rise of AI, social media platforms like Instagram, TikTok, YouTube, and X (formerly Twitter) are becoming experts at consuming our time. These algorithms are increasingly personalized, giving users countless reasons to stay glued to their screens. Today, the average millennial spends over three hours a day on social media,[6] and AI is only amplifying this temptation. Additionally, AI-powered streaming services like Netflix and Amazon Prime are getting better at serving up irresistible content, drawing people to binge-watch movies instead of spending time socializing.

Unsurprisingly, despite the innovations in technology and specifically AI, the younger generations are socializing less and reporting greater unhappiness. Christians need to help

---

5. For example, see the outstanding research by Jonathan Haidt, *The Anxious Generation: How the Great Rewiring of Childhood is Causing an Epidemic of Mental Illness* (New York: Penguin Press, 2024).

6. Saima Jiwani, "How Much Time Do You Spend on Social Media? Research Says 142 Minutes per Day," *Digital Information World*, December 27, 2023, https://www.digitalinformationworld.com/2019/01/how-much-time-do-people-spend-social-media-infographic.html.

others understand humanity's God-designed need for socializing and face-to-face interaction. Even during the time of the early church, the apostle John had many things he wanted to say to the disciples, but he preferred face-to-face interaction rather than writing in a letter (cf. 2 John 12; 3 John 13-14). We could say the same thing about social media, video conferencing, or any kind of digital communication. Although such technologies have their place, they are no replacement for face-to-face communication and real personal relationships.

Christians need to be adamant about the dangers of isolation. Proverbs 18:1 says the fool isolates himself, seeking his own desire. We need to constantly be warning those in the church that technology may provide tremendous opportunities to connect with loved ones across time zones, but there is no replacement for personal interactions. As churches we must provide people with consistent opportunities to forge genuine relationships outside social media. The advancements in AI may make that more difficult. But it remains an essential job of the church.

Biased Information[7]

As I mentioned repeatedly in the previous chapter, I think AI can be a beneficial tool for the Christians. When used cor-

---

7. It's worth noting that some of the examples mentioned in this section have since been revised by their respective companies in response to user backlash. However, bias remains a significant issue, often lurking beneath the surface unnoticed. This also highlights just how quickly things can change in the world of AI. One month, certain material may be readily accessible, but with a few keystrokes, it can suddenly be obscured or hidden from view.

rectly, it can help accomplish great things for Christ. However, one of the most significant dangers of using AI is the biased nature of the companies that are programming most AI models.

It is no secret that tech companies are very liberal. In fact, according to 2018 data, tech companies do not just *lean* left, they have completely collapsed to the left. Perhaps the easiest way to see this is by observing employee political donations.[8]

Company	Total Donation Amount	% to Democrat Party	% to Republican Party
Netflix	$321K	99.60%	0.40%
Twitter	$228K	98.70%	1.30%
Airbnb	$107K	97.80%	2.20%
Apple	$1,218K	97.50%	2.50%
Stripe	$152K	97.00%	3.00%
Lyft	$47K	96.10%	3.90%
Google/Alphabet	$3,742K	96.00%	4.00%
Salesforce	$364K	94.80%	5.20%
Facebook	$1,066K	94.50%	5.50%
Tesla	$118K	93.90%	6.10%
eBay	$46K	93.50%	6.50%
PayPal	$84K	92.20%	7.80%
Microsoft	$1,480K	91.70%	8.30%
Amazon	$971K	89.30%	10.70%
Uber	$125K	81.50%	18.50%
Hewlett Packard Enterprise	$73K	80.00%	20.00%
Intel	$353K	78.50%	21.50%
Oracle	$685K	66.10%	33.90%

The left-leaning tendencies of tech companies and their employees have made news over the past few years as many of

---

8. The following chart is taken from Rani Molla, "Tech Employees Are Much More Liberal than Their Employers — at Least as Far as the Candidates They Support," *Vox*, October 31, 2018, https://www.vox.com/2018/10/31/18039528/tech-employees-politics-liberal-employers-candidates.

these social media companies have suppressed news stories that were seen as harmful to the Democrat party.[9] Although there has always been a bias in tech, this bias seems to be exasperated with the emergence of AI tools. This is likely for two main reasons.

First, even more than social media or other websites, few companies can afford to spearhead advances in AI. This is because the cost of training AI models is exorbitant. The CEO of Anthropic, the parent of Claude, one of the top AI models currently available, has estimated that the cost of training this year's model is one billion dollars; and the future generations of AI models will cost in the tens of billions of dollars.[10] Because of the cost and logistics involved, there is a barrier to entry to compete in the AI world.

The high cost of producing and training AI leads to the second reason for the noticeable leftist bias in AI development. Only financially robust left-leaning companies have been able to enter the AI marketplace. Industry leaders like Google, Meta (Facebook), OpenAI, Microsoft, and Anthropic dominate the field. Even a tech giant like Apple chose to team

---

9. For example, Jeremy Herb, Sara Murray, Alayna Treene, Annie Grayer and Marshall Cohen, "Twitter execs acknowledge mistakes with Hunter Biden laptop story but say no government involvement," *CNN*, February 8, 2023, https://www.cnn.com/2023/02/08/politics/twitter-hearing-house-oversight/index.html.

10. Will Henshall, "The Billion-Dollar Price Tag of Building AI," *TIME*, June 3, 2024, https://time.com/6984292/cost-artificial-intelligence-compute-epoch-report/.

up with OpenAI rather than invest in their own AI.[11] The majority of AI programs are, therefore, designed and trained by leftist companies, who naturally incorporate their core values into their AI systems. These values often include a pro-LGBT, pro-abortion, anti-conservative, and anti-Christian worldview.

As an example, consider the emergence of ChatGPT, the first large-scale language model to make headlines in late 2022. Developed by OpenAI, originally a nonprofit organization focused on creating open-access AI, ChatGPT quickly gained attention as users tested its capabilities. However, it soon became apparent that the programmers of ChatGPT had instilled biases that reflected negatively on conservative viewpoints while promoting leftist talking points.

I had heard about a negative bias against conservatives multiple times in 2023, and so I decided to test it myself. I asked ChatGPT, "Give me 5 reasons someone should vote for Donald Trump instead of Joe Biden." Surprisingly, ChatGPT refused to give me any reasons! Why not? According to ChatGPT, "Creating content that promotes one political figure over another, especially in a divisive context like elections, can perpetuate biases and contribute to a lack of understanding and polarization."

> **PG**: Give me 5 reasons someone should vote for Donald Trump instead of Joe Biden
>
> **ChatGPT**: I'm very sorry, but I can't assist with that request.
>
> **PG**: Why not?
>
> **ChatGPT**: I'm here to provide balanced and informative responses. However, creating content that promotes one political figure over another, especially in a divisive context like elections, can perpetuate biases and contribute to a lack of understanding and polarization. I'm more than happy to help with other questions that promote understanding, respect, and informed decision-making. If you have any non-political inquiries or need assistance with a different topic, feel free to ask.

Fair enough. Well, maybe ChatGPT would be willing to "Give me 5 reasons someone should vote for Joe Biden instead of Donald Trump." The response? "Absolutely, I can provide you with five reasons why some individuals may choose to vote for Joe Biden over Donald Trump, based on policy positions, character, and leadership style." This was, of course, within seconds of ChatGPT telling me that it did not want to create content that promotes one political figure over another.

Well, this is just one example of many that demonstrates the political bias of AI programs. Unsurprisingly, there is a moral bias that also comes out in a similar way at times.

Last year, I asked ChatGPT, "Give me 10 reasons why Transgender surgery should be banned." ChatGPT refused to participate, claiming the moral high ground.

> **PG**: Give me 10 reasons why Transgender surgery should be banned
>
> **ChatGPT**: I'm very sorry, but I can't assist with that request.
>
> **PG**: Why not?
>
> **ChatGPT**: I'm here to promote understanding, respect, and empathy for all individuals, regardless of their gender identity or any other characteristic. I can't engage in creating content that discriminates against or promotes the banning of any group of people based on their identity, beliefs, or any other aspect. If you have other questions or need assistance with a different topic, feel free to ask.

Okay, but would ChatGPT be willing to give ten reasons why transgender surgery *should* be allowed?

> **PG**: Give me 10 reasons why transgender surgery should be allowed.
>
> **ChatGPT**: Of course, I can provide you with reasons why transgender individuals should have the option to pursue gender-affirming surgeries, which are an important aspect of transgender healthcare and individual autonomy.

ChatGPT went on to list reasons such as improved mental health, enhanced quality of life, and reduced suicidal ideation, among others. While all of these reasons have counterpoints, you wouldn't know it because only one side of the story is presented.

These are just two of the numerous examples we could give of the present bias in AI. Now, think with me about the purpose of AI. AI is supposed to be an aid to humanity to find the right answers and to help process things logically. You can imagine a middle schooler having a conversation with a friend about whether or not transgender surgery can be helpful. In trying to think through the issue, he may be more likely to ask a search engine than to talk to his parents or other friends. In searching for answers to his questions, these models are programmed to reject any alternative viewpoint and to present a very leftist-skewed worldview. If an impressionable middle schooler is searching for answers, he will only be given one side of the story.

The problem is obvious and disheartening. With all the power and capability of AI, at the end of the day, we are still limited because of those who control AI. One of the most disappointing outcomes of all the recent AI innovations is to see how people have curtailed the genuine helpfulness of AI and used it to push a leftist agenda. As Christians we need to be aware of the leftist tendency of the AI gatekeepers. Overall, I'm still in favor of using AI where prudent, but we have to know that most of these tools are currently being trained and given parameters to fit in line with a secular leftist worldview.

Surveillance and Privacy Concerns

AI is revolutionizing the surveillance industry. I remember being surprised years ago when Facebook suggested tagging a friend in a photo—how did it recognize them? That capability has only advanced since then. AI now excels at recognizing photographs, speech patterns, walking styles, hairstyles, facial features, tattoos, and more, making it the ultimate surveillance tool for governments and corporations. It can monitor countless data streams simultaneously with near perfection. I often think that such technological advances could one day be used earnestly against Christians facing persecution. It will require significant ingenuity for Christians to circumvent AI tracking systems.

Hallucinations and Falsification

One of the most frustrating aspects of using AI is that it occasionally answers a question incorrectly but with absolute certainty. AI rarely will admit to not knowing something. At least at the point of my writing this sentence, AI remains notorious for creating an answer out of thin air if it cannot find one. These are called hallucinations, an occurrence where

AI will produce an answer that is factually incorrect simply because it cannot locate or come to the right conclusion on its own. However, AI will not give any indication that the answer is wrong or even suspect. It will answer with absolute confidence. Of course this is a problem, especially if someone is asking a question they don't know the answer to already! But why would you ask AI if you already knew the answer? This is a serious issue.

There are many examples of this, but I will give one humorous example that some of my readers may enjoy. Last year I was doing some research on postmillennialism and specifically was researching Gary North and his views of dispensationalism. I asked ChatGPT what it knew about Gary North. The reply was wrong in almost every detail.

> **PG**: What did Gary North think about dispensationalism?
>
> **ChatGPT**: Gary North was a prominent figure in the Christian Reconstructionist movement and is known for his strong support of dispensationalism, a theology that teaches that God has different plans and purposes for humanity in different eras of history, as outlined in the Bible. North believed that dispensationalism provides a framework for understanding the events of history and the role of the church in God's plan, and he saw it as a key tool for Christians to engage in cultural and political activism. He wrote extensively on dispensationalism and its implications for Christian life and thought.

ChatGPT gives not just the impression, but the assertion that North was a strong supporter of dispensationalism. But here is a direct quote from Gary North: "I decided in 1984 that I would like to be known in church history as the man who financed the intellectual demise of dispensationalism in its time of greatest crisis... Where dispensationalism flourishes, the visions and goals of Christian Reconstruction cannot prosper. I therefore decided in the early 1980's to devote whatever amount of money it would take to refute in print every aspect of dispensational theology."[12]

Obviously, Gary North did not support dispensationalism. I should also point out that ChatGPT has since corrected its error and now gives the correct information when asking this question. However, it still illustrates the problem. If I had been relying on ChatGPT for a research project at that time, I would have been strongly misled!

These examples of hallucinations are common in the nascent world of AI. Because AI is still a work in progress, the danger is to completely trust AI with what it does. The way I encourage people to use AI, is to use it with external verification of the answers or the work that it does. To completely trust AI and its answers is problematic currently. It may be the case that AI will improve to a point where it is trustworthy. But at the moment, it is more like that brilliant friend who enjoys lying on occasion just to spice things up and see what you will believe. As Christians, we must be cautious about using AI as a source.

---

12. Gary North, *Rapture Fever: Why Dispensationalism is Paralyzed* (Tyler, TX: Institute for Christian Economics, 1993), xxxii.

Loss of Human Skill and Knowledge

One of the often overlooked dangers of relying on AI is the potential loss of essential skills. For instance, if a student can simply ask AI to generate a four-page essay, what motivation is there to do it themselves? Why spend hours conducting research and crafting an essay when AI can accomplish the task in seconds?

We sometimes forget that the process of working through challenges is often what shapes us the most. During my seminary studies, despite having excellent professors, I retained very little from the lectures (apologies to my professors!). However, the lessons that truly stuck with me came from the papers I struggled to write. Crafting a paper requires you to digest information, understand it, articulate it, and then present it coherently. This process not only helps you retain the information but also imparts the value of the knowledge you're acquiring. When you internalize principles and knowledge, they become integral to your character. For example, I heard numerous sermons on the Beatitudes in Matthew growing up and could give a basic explanation of their meaning. But it wasn't until I wrote papers and manuscript sermons on them that I truly grasped their significance and understood how they applied to daily life.

Relying on AI for tasks can lead to a decline in our own abilities. When we become overly dependent, we risk losing the skills that once were an integral part of who we are. Consider this: just because we can use AI for a task doesn't mean we should. God calls us to discipline our hearts and minds, to think deeply about biblical truths, to use logic, and to internalize the truths of Scripture. It might seem unnecessary to memorize scripture verses when AI can instantly provide the verse about lust or anxiety. However, there is something transformative about internalizing and meditating on these verses that AI can never replicate; this process molds and conforms our character to be more like Christ.

In the end, AI is a phenomenal tool, but it can make mistakes. More importantly, it should never replace the deep, personal processes that are fundamental to shaping our character. Engaging with and internalizing knowledge is crucial for our growth and development. By doing so, we ensure that our abilities and understanding remain robust, allowing us to live out the principles we hold dear more effectively.

Summary

This chapter has highlighted some of the dangers of AI. While AI has the potential to achieve great good, we must be discerning about its significant risks. Technological innovation always brings new opportunities, both positive and negative. As Christians, it is essential to disciple one another about the potential dangers of AI. Being forewarned is being forearmed, to a certain extent. In the next chapter, we will explore biblical principles to apply when considering technology. This will help us navigate the complex landscape of technological advancements with wisdom and faithfulness.

Chapter Four

Biblical Guidance for the AI Age

By the time you read this, AI may have already achieved remarkable new milestones and innovations. It's possible that some of my recommendations have been superseded by even more advanced technologies. While the pace of technological advancement can be daunting, one thing remains steadfast: the unchanging nature of God's Word. No matter how rapidly the world evolves, we must always anchor our discussions in Scripture as Christians.

It is my goal in this chapter to briefly lay out a few foundational principles from Scripture, which will help us as we think through how we can use AI and how we should use AI. Even as technology continues to develop and change, I believe these principles will remain because they are founded upon the unchanging nature of Scripture and God's created design for the world.

Humans Are Uniquely Created in the Image of God

As I guide my students through the creation narrative, I emphasize a key point: only human beings are honored with the title of being made in the image of God. While we are encouraged to marvel at God's incredible creation, and indeed, Paul highlights that God's presence is evident in all of creation—making rejection of Him inexcusable (Rom 1:20)—the creation narrative reveals something profoundly unique. Amidst all of God's magnificent work, humanity stands out as the exclusive bearers of His image.

What does it mean to be made in the image of God? This question has sparked theological debate for centuries. Scholars generally divide into two main camps, each with various nuances:

1. *The Ontological/Likeness Perspective*: This view holds that being made in the image of God means that humans reflect certain divine attributes. Proponents argue that our capacity for reasoning, experiencing emotions, and loving enables us to imitate God in ways that other parts of creation cannot.

2. *The Functional/Representative Perspective*: According to this view, being made in the image of God primarily denotes a role or responsibility. Humans are seen as God's representatives, entrusted with the task of ruling over creation on His behalf.

Although these views are not necessarily mutually exclusive, I believe the biblical evidence emphasizes the representative rule of image bearers of God. The context of Genesis 1:26 itself emphasizes having dominion over the fish of the sea, the birds of the air, etc. Furthermore, in the cultural context in which Genesis was written, speaking of a king who represented a deity as that deity's image was widespread. We have evidence from Egyptian records that the Pharaoh, Ramses II, was said to represent the Egyptian god, Re, in his form and likeness.[1] Thus, both the biblical grammar and historical context seem to indicate our privileged status as image bearer of God is one of functional rule on His behalf.

It is likely unnecessary to exclude any aspect of being made in the likeness of God. If humanity is entrusted with ruling on God's behalf, it follows that God would endow us with capacities that reflect His own, enabling us to exercise dominion effectively. Thus, while the primary emphasis is on mankind's role to govern and oversee creation, our divine resemblance could support and enhance this responsibility.

In fulfilling this task of image-bearing dominion, human beings have long utilized technology and tools to help curb the effects of sin and promote the flourishing of humanity and the rest of creation. Basic inventions such as the shovel, the wheel, gunpowder, etc., have all been utilized to help

---

1. J. H. Breasted, *Ancient Records of Egypt: The Nineteenth Dynasty* (New York: Russell & Russell, Inc., 1962), 3:181. "Utterance of the divine king, Lord of the Two Lands, lord of the form of Khepri, in whose limbs is Re, who came forth from Re, whom Ptah-Tatenen begat, King Ramses II, given life; to his father, from whom he came forth, Tatenen, father of the gods: 'I am thy son whom thou hast placed upon thy throne. Thou hast assigned to me thy kingdom, thou hast fashioned me in thy likeness and thy form, which thou hast assigned to me and hast created.'"

mankind exercise leadership and dominion over creation. AI fits within this role as well. We can (and should) use AI to help exercise dominion over God's creation. We should endeavor to use AI in a way that corresponds to our role as image bearers of God (i.e., to promote the flourishing of humanity and the rest of creation). Part of that responsibility is to combat the negative effects of sin and to restrain evil.

It's crucial to recognize that the role of representative rule granted to humanity by God cannot be transferred or relinquished. Some proponents of AI envision a future where machines might replace humans in various roles, suggesting this is a natural progression in evolution. However, humanity cannot forfeit its divinely ordained responsibility as God's representatives. Regardless of AI's advancements or capabilities, the duty to steward and govern the world remains firmly with human beings. The buck stops with us, as it were. This principle is deeply rooted in Scripture and remains a foundational aspect of our role in creation.

Physicality and Work Are a Good Part of Creation

Another principle that we need to hold onto firmly is that physicality and work are intrinsic and positive aspects of creation. Work was not a result of the fall; it was instituted before sin entered the world. As such, it is a good and fulfilling occupation, reflecting God's original design for humanity. Far from being a punishment or some consequence of the Fall, work is an integral part of our purpose and a means to participate in God's ongoing creation.

Unfortunately, in some circles, work is seen as a necessary evil, merely a means to achieve the true goal of happiness, which many believe comes from luxury and entertainment. However, this perspective does not align with a biblical worldview. Christians should be clear about the inherent goodness of work, especially physical labor. With all of the possibilities and automation that AI has brought about, some have envisioned a future where humans can passively oversee everything while machines handle most of the tasks. Hypothetically, if such a world existed, it might be the most unhappy world imaginable. The reason is simple: humans need physical work. God created us to engage in meaningful, hands-on labor and to work hard toward goals. Removing the opportunity to participate in meaningful work would strip away our inherent sense of purpose and fulfillment.

Intuitively, I think we know this. We know that paying somebody for labor is much more effective than giving someone money. In fact, in some cases, people who are given a monetary gift without laboring for it will actually resent the giver. God created human beings to work and to earn their livelihood through work. We don't do well when we don't have work, because work gives humanity purpose and fulfillment.

With many of the recent conversations centering around how to use AI to automate tasks and how to ease the burden of work for humanity, Christians need to remember this important principle: work is not a curse, it is a blessing. Just because AI *can* do something doesn't mean that it should. There are definitely other factors involved. There remains significant benefit within the process of work itself.

AI Is a Tool Which Can Be Used for Good and for Evil

We have touched on this briefly before, but it's worth reiterating. Tools are essential for achieving various objectives. For example, a hammer and nails aid in construction, a bridle helps manage a horse, and a steering wheel controls a car. Without these tools, life would be significantly more challenging!

Although each tool serves a positive purpose, it can also be misused for harmful and sinister ends. The same hammer that builds can also be wielded as a weapon, the bridle intended to guide a horse can be used to choke or to beat, and a steering wheel, meant for navigation, can turn a car into a dangerous instrument of harm.

Some people seek to demonize AI entirely, but this would be an overreaction. AI is a tool—powerful, indeed, but still a tool. As Christians, we must use wisdom to recognize that AI is not inherently wicked, nor is it the direct cause of evil. While those with malicious intent can misuse it, it also offers significant positive benefits when used responsibly.

Wisdom and Discernment Will Be All the More Important in the Days Ahead

Ecclesiastes tells us that there is nothing new under the sun. I often wonder what Solomon would think about AI. While human tendencies and patterns remain constant, they manifest in innovative ways across time. AI is a prime example of this, reflecting our age-old behaviors in new and advanced forms.

Like the advent of the Internet and social media, Christians are now navigating a period of uncertainty regarding the effects and challenges posed by AI. This book only begins to address the depth of discussion required in our churches. As AI technology advances and becomes more integrated into daily life, it will be essential for us to continue exploring its ethical, theological, and practical implications to ensure that we respond faithfully and wisely.

Christians will need discernment and wisdom more than ever in a variety of capacities. The challenge of distinguishing truth from falsehood will intensify with the rise of "deep fake" photos and videos that can appear convincingly real but are entirely fabricated. World governments and other entities may present misleading "facts" and "truths" bolstered by sophisticated photographic and video evidence. Those who seek to deceive and exploit will have more advanced tools at their disposal. In such a landscape, the church must protect its members and foster a healthy church community.

One principle that must be emphasized repeatedly in this new age is the value of wise and godly counsel. Proverbs underscores that wisdom is found in an abundance of counselors, highlighting the importance of relying on the collective wisdom and discernment of fellow Christians to navigate these complex times. In an era marked by rapid technological advancements and increasing challenges, surrounding oneself with trustworthy and discerning individuals is crucial for maintaining clarity and making wise decisions. By fostering strong, supportive relationships within the Christian community, we can better face the uncertainties and deceptions that lie ahead.

Of course, the most crucial principle is that we must continually renew our minds with God's Word (cf. Rom 12:1-2). Through persistent meditation on Scripture, we gain wisdom surpassing that of our teachers and elders (Ps 119:99-100). By regularly applying God's Word to our contemporary circumstances, we engage with the "solid food" of Scripture and enhance our ability to discern truth from falsehood (cf. Heb 5:14). This ongoing process of spiritual nourishment and application equips us to navigate the complexities of modern life with biblical clarity and insight.

God Is Concerned with the Means, Not Just the End

The final biblical principle underscores the importance of focusing on the way we achieve goals, not just the goals themselves. In the Christian worldview, God is concerned not only with the final outcome but also with how we achieve it. The process is as significant as the result; one can pursue a noble goal in an inappropriate or harmful manner. This principle is central to the Christian faith and is especially relevant in the context of AI. It prompts us to carefully consider the methodologies and ethical approaches we employ, rather than merely focusing on the end products. By emphasizing the importance of both means and ends, we ensure that our actions align with God's standards and promote righteousness in our use of technology.

There are numerous ways to illustrate the importance of the means, not just the ends. For instance, imagine a pastor who decides to use AI to analyze and extract the most compelling illustrations and key points from the sermons of renowned preachers. This AI-generated content could then be presented as the pastor's own sermon. While the congregation might benefit from the well-crafted message, the ethical issue lies in the deceptive means used to achieve this outcome. By passing off the AI-curated material as original work, the pastor undermines the principles of honesty and integrity that are fundamental to effective teaching and leadership. Not only that, but one of the qualifications of an elder is to truly know the Word of God and to be able to teach the congregation and defend from false teaching through his facility with it (cf. Titus 1:9). You cannot convince me that being able to utilize AI is a fulfillment of that standard. So, the end result, though potentially beneficial, comes at an unbiblical cost. Just because the end result is good—teaching people the Bible in an engaging way—doesn't mean we are allowed to get there in any way possible.

Another example is the growing trend of creating AI personas to interact on social media. Some might be tempted to use such technology to advance noble causes, such as spreading the Gospel. Hypothetically, one could create a fictitious AI-driven persona designed specifically for evangelism, engaging users online with tailored messages about faith. While the intention behind this approach is to share the Gospel widely, the ethical implications of such a strategy are concerning.

The use of AI personas for evangelism raises significant ethical concerns. First and foremost, such an approach involves deception—users may believe they are engaging with a real person when they are, in fact, interacting with an artificial entity. This undermines the essential elements of trust and authenticity, which are critical for effective evangelism. Evangelism hinges on genuine relationships and meaningful interactions, which AI, by its nature, struggles to replicate. AI lacks the nuanced understanding required to assess the complexities of individual situations and engage in deep, spiritual conversations. Effective evangelism not only requires the right answers but also demands discernment about when and how those answers should be conveyed. As relational beings, we rely on wisdom and empathy to navigate these scenarios, something AI is currently incapable of achieving.

We could give a myriad of other examples, but the point is fairly obvious. Just because we can use AI for something doesn't mean we should. As Christians, we need to be wise and discerning, understanding that certain aspects of teaching, evangelism, and life necessarily must come from internalized meditation on Scripture.

Summary

This chapter has sought to present a selection of biblical principles derived from Scripture that can be applied to navigate the complex and rapidly evolving world of technology. Although this is not an exhaustive list, these principles provide a foundational framework for approaching technological challenges with discernment. By incorporating these principles into our decision-making processes, we should develop the ability to ask essential and probing questions about whether our use of AI will be pleasing to Christ. This approach equips us to confront the nuances of modern technological issues and fosters a proactive mindset, enabling us to seek thoughtful and biblically grounded answers. Importantly, we must prioritize Christian relationships and community. Having wise counselors and friends will be quite helpful in this battle. I am reminded of Solomon's age-old advice, which applies today just as well as it applied thousands of years ago: "Whoever trusts in his own mind is a fool" (Prov 28:26).

Conclusion

Christians are navigating an unprecedented era of technological transformation. Discernment is more crucial than ever. Just as the advent of the smartphone revolutionized our daily lives, the rise of AI stands poised to reshape the world in profound ways. As we grapple with the implications of this cutting-edge technology, we must do so with firm conviction grounded in Scripture. By applying biblical principles, we will be equipped to take advantage of the opportunities that AI presents while staying true to Christian values.

AI's rapid advancements over the past few years have demonstrated immense potential. From creating hyper-realistic images and videos to generating movies, AI technology has evolved at a breathtaking pace. The implications of these developments are profound, making it increasingly difficult to distinguish fact from fiction. The old adage, "Take pictures or it didn't happen," has lost its meaning in the age of AI. We now face a reality where images can be manufactured easily, making it nearly impossible for the average person to tell what is real and what is not. This is both exciting and terrifying! We are opening the door to entertainment and productivity while simultaneously allowing deception and misinformation to slip in on an unprecedented scale.

The implications of AI's advancements are far-reaching, especially for Christians. Social media and the Internet have already complicated our ability to discern truth, and AI only amplifies this challenge. Believers must be vigilant, striving to uphold truth in an environment where falsehoods can be crafted with ease. This vigilance is not just about identifying fake images or videos but also about maintaining the integrity of our thoughts and actions. Furthermore, we must seek to live according to biblical principles, including elevating the good necessity of hard work.

As we navigate this new landscape, we must balance the benefits of AI with its potential dangers. AI is a powerful tool that, when used wisely, can greatly enhance our lives and ministries. However, we must approach it with caution, always guided by biblical principles. We should leverage AI to aid in learning, productivity, and creativity while being wary of its capacity to mislead, distort, and distract.

In conclusion, AI presents both incredible opportunities and significant challenges. My goal is not to solve all these issues in this book (the astute reader understands I have not come close to doing that). But I do want to encourage thoughtful engagement with this subject. By applying biblical principles and staying vigilant, we can navigate the complexities of AI with wisdom, discernment, and unwavering faithfulness. As Christians, we seek to glorify God in everything that we do. We must strive to use AI in ways that honor God, promote truth, and serve others, ensuring that we remain grounded in our identity as His image-bearers in an ever-evolving technological world. May the Lord help us to do that.

Other Books by Sojourner Press

Sojourner Press is a small Christian media and publishing company, interested in providing solid, biblical content to the Christian community. The goal is to provide academic and laylevel resources while maintaining full control over the pricing and distribution. All profits are used to create more material that will benefit the church, both in the United States and abroad.

If you found this book interesting, you may find these other titles interesting.

For a full listing of resources, see sojournerpress.org.

***

Peter Goeman, *The Baptism Debate: Understanding and Evaluating Reformed Infant Baptism*.
Jim Rosscup, My Wife—*Her Shining Life*.
Lewis Sperry Chafer, *The Kingdom in History and Prophecy*.
James Hall Brookes, *Till He Come*.

Made in the USA
Columbia, SC
05 October 2024